The Chronolo

MW00440347

Real Blac...

by Ankh West

Amen Ra Squad

Voices of Fire Press
Voices of Fire Press purpose is to educate,
inform, and the persuade the reader towards
adopting a sound approach to our history and
information with the overall goal of redefining
what it means to be conscious.

A Chronology of Human Evolution- Real Black Atheism Explained

@2016 Written by Ankh West

Published by Voices of Fire Press 2016

Voices of Fire a Press is Company of United Consortium, LLC

St. Louis, Missouri 63124

Inspired by Dr. -Dr. Yosef Ben Jochannan

"Chronology of the Bible"

Introduction to Real Black Atheism:

My interest in Real Black Atheism was magnified after joining the group the Real Black Atheist of Atlanta. The team was an excellent platform for individuals interested in traditional African culture as a means of networking, sharing, and educating each other on African Centered Atheism. Around that time, I also joined The Black Atheist Group thinking the group would also allow me to connect with people who shared my same beliefs, however as I became more acquainted with some of the group members, it becomes more and more apparent that like European Atheist, Black Atheist also refused to acknowledge that nature was the foundation of African Spiritual Systems. Like European Atheist, Black Atheist declined to recognize the ancient wisdom of our culture and the reality that God was a manifestation of nature and comes out in the minds of men.

In many ways The Black Atheist group was worse than some of the white Atheist groups because they not only rejected nature, they also denied the validity of RBA. Of course, the failure to acknowledge the validity of RBA as an African paradigm is directly correlated with religious indoctrination in the community. As a result, I was forced to deal with the scrutiny that was equally distributed but what's apparent is both atheists, and Theist regards RBA as an invalid premise and rejects the notion that there are differences between an Atheism and a Real Black Atheism. Given this, I decided to publish this Booklet to explain the meaning of RBA and re-establish the importance of the restoration of our culture.

Before we look at the differences between RBA and BA, we must first look at the etymology of the word Atheist, which generally, describes a person "without god" or one that "denies God." (1).

An atheist does not acknowledge the differences between black a white cultures, they do not support traditional systems, and most of the time they correlate African agency as pseudo. Therefore, using the European paradigm, an Atheist is simply, "without a god." This is important when addressing the behavior of Black Atheism as a separate discipline from white atheism, because as we can see atheism does not deal with culture, yet the reality of being a black atheist is apparent in the community.

In fact, just like the dominant society Black atheist are ignored, marginalized, and left-out of positions of power. Black Atheist are left to develop its paradigm called Black Atheism. However, where Black atheism fails, RBA succeeds. Manly, because Black Atheist refuses to acknowledge culture and opt for skin color politics. Therefore, a black Atheist is simply someone who has melanin and who rejects the gods. Specifically, a black atheist does not know the history of African Spiritual Systems.

A Black Atheist does not want to learn about Africa and thinks that all things in Africa are stupid, outdated, or barbaric. A Black Atheist does not know or understand African cultures had their spiritual systems before the word god, and they lack an understanding of what separates a black Atheist from a Real black atheist. Because they refuse to acknowledge ancient African systems that our ancestors used to create concepts found in nature.

Even though we can trace the origins of Atheism to Greeks, the term is attributed to 14th century Europeans. Despite this, the father of Atheism is the Greek philosopher named Thales. The Greeks played a significant role in the development of Atheisms, and many of them learned from Ancient Kemet. Thales was one of the first philosophers to travel to Kemet to study science. He was born in Ionia (present-day Turkey) around 624 BC at a time when it was a Greek territory. And according to ancestor George G.M. James in Stolen Legacy: Greek Philosophy is Stolen Egyptian Philosophy, "... history makes it clear that the surrounding neighbors of Egypt had all become familiar with the teachings of Egyptian Mysteries many centuries before the Athenians...because philosophy was something foreign and unknown to them." (2). James notes that the territory was an Egyptian mystery school "stronghold." So, not only did Thales learn from the priests in Kemet, he grew up under their influence.

After studying with the priests, Thales opened schools in Ionia to pass on the knowledge he had learned from the African priests. His students included Socrates, Aristotle, and Plato, and the African expertise they all learned was called Voodoo. "Know Thyself" is ancient Kemet wisdom. Thales urged his students repeatedly to "Know Thyself." It was part of the philosophy that he learned from the African priests and passed on to his students. But, he also said, "The most difficult thing in life is to know" yourself" probably because it was difficult for Thales to understand his nature. When the ancient Greek leaders heard what Thales was teaching, they called him an atheist and claimed that his information clashed with their knowledge of the Greek gods. Thales never called himself an atheist. (3)

In fact, he is credited with saying, "everything is full of gods." (4) And though Thales is said to have paid little attention to the Greek gods, he was deeply influenced by African spiritual and philosophical principles. Still, the Greeks labeled Thales, an atheist. James writes in Stolen Legacy that Thales and his students were persecuted and driven from Greece: "The indictment and prosecution of Greek philosophers is a circumstance which is familiar to us all. Several philosophers, one after another, were indicted by the Athenian Government, on the common charge of introducing strange divinities." (5). And that was because the political system of ancient Greece cherished Greek gods.

Any opposition to those gods meant opposition to that political system. Opposition to the political system made one an atheist. It's a concept that is entirely different from what we understand as atheism today. But, hear me and understand that whenever we learn traditional African social system, we become an Atheist. Thereby anyone who rejects state-sponsored became an enemy of the state and by default labeled an atheist.

This is essential to the concept that gave rise to the Real Black Atheist. RBA reject the gods of Greece, of Christians, and of Islam. It lifts African tradition and focuses on nature, food, water, and shelter.

A Real Black Atheist does not follow false deities because (Africans) invented the idea and the concept of God. However, what is different about a Real Black Atheist is that we understand that when our ancestors talked about God, they were merely talking about nature, the cosmos, and the universe. They were showing reverence to the ancestors, giving praise and admiration to the forces, and they understood the power of nature. This had nothing to do with praying, spook-ism or invisible things.

Real Black Atheism, therefore, goes back to the womb of Mother Africa where culture and civilization were formed. The Black Atheists should never lose their African religion because it was the African culture that made the Greek philosophers turn against the Greek gods; it is what made the Greeks call them atheists. A Real Black Atheist does not leave religion just to praise European culture. A Real Black Atheist does not refuse to read and study the great works of Dr. Ben, or Cheik Anta Diop or John Henrik Clarke. A Real Black Atheist understands the importance of today and the vision of the ancestors.

"Africans had a religion in Ancient Egypt there were two types of religion in the world: supernatural religions or theistic religions. And then you have nature religions or atheistic religions. A lot of people have an idea that if a man is an atheist, that is: if he doesn't believe in a personal god, he's an irreligious person. That's not true. The Africans had an atheistic type of religion, and they were a deeply religious people. Because they believed that man not only has a body, but he also has a mind, soul or spirit...if you went to the temple in ancient Egypt, and you told the priest that you wanted Horus the Egyptian Christ to save your soul, he would tell you not to waste his time. Horus, Osiris, Isis or nobody else was going to save your soul, except you."-Dr. John G. Jackson

Now, that we have looked at Real Black Atheism, let's take a quick look at traditional African systems to see what Thales and other early Greek philosophers learned in Kemet. First of all, primary researchers, scholars, and elders such as Dr. Yosef Ben Jochannan said that Voodoo had nothing to do with witchcraft. In his book, African Origins of Western Religion, Dr. Ben writes that "voodoo, voudou, or vaudoux comes from the Dahomey in West Africa where it means a genius or protective spirits." (6) Milo Rigaud writes in his book, "Secrets of Voodoo," that "vo" means introspection" and "du" means into the unknown (into the mystery)." (7)

What that all means is that Voodoo is a creative force: it is a belief system joined in union with religion? They are compiled together, always to be learned together, and to take advantage of the creative force. When we learn about that energy, we get math, science, arts, etc. The math, science, and arts are the outward expression of who we are inside. The Africans had mastered religion. And religion was the explanation of Science. We didn't separate religion from science. Europeans separated the two. When we take an inward look at self, we start to unravel life, which is the mystery. When we begin to explain that magic, we become a genius. So, Voodoo means that whenever you take an inward look at self, you become a genius.

When the Africans of Kemet mastered Voodoo,

they built great pyramids, magnificent temples, great universities and great shrines to their ancestors. That is why Thales encouraged his students to "Know Thyself." Therefore, when you study self and grow to know yourself, you will build great civilizations based off of math, science, and agriculture. It is already inside of you. It does not come out of the air. Knowing self-does not come from spook-ism; it comes from hard work. Unraveling that mystery makes you a genius. If you're still wondering which god or gods, this Real Black Atheist believes in then relax, I don't believe in a god because God and Nature are one in the same. I study to know nature. This is what our ancestors called MuKulu. MuKulu is an African word and is the oldest etymology of the word god. MuKulu is intended to represent distance. The highest or height in the sky or highest achievement that comes along with age.

MuKulu is also intended to represent water which is linked to the sky in the form of rain that brings sustenance and prosperity to the community. (8) When I use the word god, what I'm expressing is MuKulu so that it speaks of someone who invested his or her time in building up and providing for the community. MuKulu is what our ancestors invented. It was their god concept. The secondary meaning for MuKulu was a wise old elder, a great king or great queen or it was an ancestor.

A child could not be MuKulu because the word expresses distance or age. Those ancient Africans, our ancestors, said MuKulu before there was colonialism; before there were Greeks; before there were Persians; and before the Romans, we knew God/Nature as a wise old elder or as the earliest representative in their country. This is how the original African people who invented the word god and its concept intended for them to be represented. Not surprisingly, the oldest representations of god in art form are the sky portrayed as a woman.

Women were at the center of the African culture because she gave birth to the community (children, science, writing, math, art, etc.) and the workforce. Other cultures copied and borrowed their gods from Africans who made statues worshiping themselves (us). Therefore, A Real Black Atheist would never want to go against the ancestors because they built up the community. A Real Black Atheist would never want to be against the sky that brings life in the form of rain. A Real Black Atheist would never want to go against the wise old elder because they would never learn.

I am a Real Black Atheist, and the Mukulu are my ancestors. I know nature because nature is in me. Simply put, I am an aspect and the essence of nature (9) having a human experience.

Nature is the elements, the forces, the powers and the energy that creates. Now tell me who you are? Tell me what you know and not what you believe. But before you speak, remember that first-hand research kills the conversation every time.

"Read books about yourself. There are millions of them. Let no one tell you to the contrary. Read about religious doctrines and theories. Don't let anyone tell you: 'this is it.' And there is no more... Read, read, read. Read about your people in the past, the present, the future."

-Dr. Yosef Ben Jochannan

I am grateful to my beautiful mother, my wife, and kids as they support me in the journey towards achieving my dreams and I'm appreciative of my fellow Amen Ra Squad Members & Affiliates in their support, inspiration, and continued encouragement towards knowledge of self. Black African *Power- Real*

Black Atheism on the rise.

Evolution Explained

Evolution is a concept that goes hand in hand with the idea of life itself. All life that we see around us, all life that exists has undergone millions and millions of years of evolution. It can be described as the change in heritable characteristics of organisms over successive generations. All life that exists in this world has undergone evolution, to be able to survive and thrive in the changing environment. Human evolution follows the same rules and patterns.

Let's break down evolution and see what the key concepts are. The basic idea of evolution is that all life present on this Earth, share a common ancestor. Through diversity we observe in fossils and around us today, the common ancestor evolved into different forms of living organism that we see today. So how does evolution work? Well, we all inherit genes from our previous generation; but as generations expand throughout time, mutations occur in that genetic code, and when further generations inherit that mutated genetic structure, additional mutations come up in the following generations. By then, the original genetic structure and the one from the last generation observed develops new code that has evolved through mutation process.

Then there is the mechanism of genetic drift, which is one of the mechanism of evolution that is heavily centered on chance. This depends entirely on the concept of "strength in numbers", there is one generation that is leaving behind a large number of descendants, the following generations, therefore, will have a greater number of those genes; these genes may or may not be better or healthier, but since they are greater in number, they have a better chance of survival. This mechanism is called Genetic Drift. Then we have the Mechanism of Natural Selection which is similar to Genetic Drift, but different because in this mechanism only the stronger and better genes survive.

The surviving generation did so only because it was better and stronger than the rest of the generations, and so the following generations will have genes from this generation. In a way, nature favors those genes that are strong and able to survive, and thus only their generations will make it further because they will have similar characteristics.

How does this work in human evolution? This works the same way in human evolution, the same mechanism applies. Mutations come up in following generations; the same genetic code is passed on to further generations but with few variations. In human evolution, genetic drift existed as the genetic code of only those generations that survived whose numbers were significant. And finally, through natural selection, just those characteristics that favored the environment made it through. In prehistoric time, the humans who could hunt, kill and defend themselves survived as compared to those who died as they could not do so. The following generations had the survivor's genetic code, as that was the only code left. By these three mechanisms, we see how human evolution has made it.

In the following pages, we will document the origins of the fossil record from different species that shaped our evolutionary process and significance of Africa in the development of our current species called Homo Sapien Sapien or Modern Man. The following fossil records span from 7 million years to present day. The First set of Fossil records is of hominids called Australopiticus, while the second set is for Homo or the Human family. The order of the fossil record is not necessary in order but organized in a way to help the reader see the process of evolution. Africa was the center of human history that shaped the human family, and although there have been many discoveries on the continent, its clear our history goes beyond the bible and any African looking to find himself, should first start with Africa.

"Understanding Evolution." Understanding Evolution. N.p., n.d. Web. 05 Dec.

Fossil Records
(This order does not necessarily represent the sequential order)

7 Million Years

Sahelanthropis tchadensis-Discovered in Chad in 2001-Nickname Toumai. The fossil record includes a complete Skull. It's unclear if this species was Bipedal.

6 Million Years

Orrin tugenesis-Discovered in Western Kenya in 2000-nicknamed "tugenesis" because it was found in the Tugen Hills of Kenya. The fossil record includes fragments of the arm, thigh bones, lower jaws, and teeth.

4.4 Million Years

Ardipithecus ramidus-Discovered in the Middle of the Awash Valley in Ethiopia between 1992-1993 Nicknamed "Ardi" who was believed 3 '11''. Ramidus has a small brain, bipedal, and is believed to have lived in a wooded forest terrain.

4.2 to 3.9 Million Years

Australopithecus anamensis-Discovered in Kenya and Ethiopia in 1964 with over 100 fossils totaling 20 individuals. The fossils include teeth that resembled primitive animals and a skull that has human features.

3.9 Million Years

Australopithecus afarensis-Discovered in Adar region of Ethiopia in 1964. Afarensis is one of the most famous fossil records in history known as "Lucy." She had Human-like hands and teeth with a small physical statue and was believed to be bipedal. Lucy is also called "Dinkinesh" meaning "you are marvelous" in Ethiopian.

3.5 Million Years

Kenyanthropus platyop-Discovered in Lake Turkana Kenya in 1999. The name Kenanthropus was given, due to the significance of Kenya in Human Evolution.

3 to 2 Million Year

Australopithecus africanus-Discovered in South Africa in 1924. This species is believed to be the pre-human ancestor with similar features to Afarensis and has a human-like a skull.

2.5 Million Years

Australopithecus garhi-Discovered in Bouri, Ethiopia in 1998. This Bipedal Australopithecus had large rear teeth, a primitive skull, lived mostly off a plant-based diet and possibly some meat.

1.78 to 1.95 Million Years

Australopithecus sediba-Discovered in Malpa South Africa in 2002. The skull is believed to be the transition between Africanus and Homo genus family.

2.6 to 2.3 Million Years

Australopithecus aethiopicus-(Black Skull)because mineral uptake during fossilization gave the specimen a blue-black color. Discovered in Southern Ethiopia west of the Omo River in 1986. The skull appears to shows a mixture of both primitive and derived features.

2 to 1.5 Million Years

Australopithecus robustus-Discovered in South Africa in 1938.The evidence suggests that this species diet was mostly course, rough foods, and teeth indicate the food needed a lot of chewing.

2.1 to 1.1 Million Years

Australopithecus Boise-Discovered in Tanzania in 1959 with a Nickname "Nutcracker" man. Boise is thought to have been related to modern humans, but its large molars suggest that it may be related to robustus
Homo-Genus Family (Human Family)

2.4 to 1.5 Million

Homo-habilis-Discovered in Tanzania in 1959. Also known as "Handy Man" due to the many tools found with its remains. Homo Habilis had a larger brain and is believed the be an earlier ancestor.

1.8 Million Years

Homo-geogicus-Discovered in Dmansi Georgia 1999-2002 is thought to be an intermediary between H. Habilis and H. erectus.

1.8 Million to 300,000 Years

Homo-erectus- Discovered first in 1891 in Asia. Also, know as Upright Man due to his upright statue and robust features. Some important fossils are Turkana Boy and Peking Man.

1.89 Million to 143,000 Years

Homo-egaster-Lived in South Africa but believed to share some common features with homo-erectus.

78,000 Years

Homo-antecessor-Discovered in a Spanish cave site in Atapuerca making it the oldest European Fossil. Named in 1977. Its facial features appear modern, but teeth, forehead, and brow ridge are primitive.

500,000 Years

Homo- "Archaic" Saipan heidelbergensis- The Term covers a group of skulls that have similar makeup as H.erectus and modern man making it an Archaic Human.

230,0000-30,000 Years
Homo-neatherdalsenis- Discovered in Europe in 1929. The Neanderthal co-existed with modern man for long periods of time. However, modern man does not come from the Neanderthals.

38,000-18,000 Years
Homo-floresiensis-Discovered in Indonesian Island of Flores in 2003. The complete fossil is an adult female. Floresiensis is also known as the "hobbit."

195,000 Years Ago to Present
Homo Sapiens Sapiens-Modern Humans-Originated from the Omo Valley in Africa. Modern Man is the last surviving species in the homo-genius family. Modern man is believed to be the most invasive species in the history of the planet.

---

10. Foley, Jim. "Hominid Species." Hominid Species. N.p., n.d. Web. 05 Dec. 2016.

12. Smithsonian's National Museum of Natural History. "Human Fossils | The Smithsonian Institution's Human Origins Program." Human Fossils | The Smithsonian Institution's Human Origins Program. N.p., 01 Mar. 2010. Web. 05 Dec. 2016.

13. Cookson,Https://www.facebook.com/BradshawFoundation Clive. "Discovery of Earliest Homo Sapien Skulls Backs 'Out of Africa' Theory." Bradshaw Foundation. N.p., n.d. Web. 05 Dec. 2016

Frequently Asked Questions

Q) What does Biological evolution mean?
A) Biological evolution merely is understood to be, descent with modification.

Q) Does human Evolution explain how humans came to earth?
A) No, it only explains how humans have evolved on planet earth.

Q) Does Evolution explain where life on earth originated?
A) Evolution does not explain where life on earth comes from it only seeks to explain how life evolves once it gets to here

Q) Who claims that humans come from monkeys?
A) People who have not taken the time to study and understand human evolution. The claims, for the most part, start with the religious zealots and creationist who fight against science with ignorance, not facts.

Q) What is a primate?
A) Any of various mammals of the order Primates, which consists of the lemurs, lorises, tarsiers, New World monkeys, Old World monkeys, and apes including humans, and is characterized by nails on the hands and feet, a short snout, and a large brain.

Q) Are human's mammals?
A) Yes

Q) What makes humans mammals
A) The fact that humans breathe air has a backbone and grows hair at some point during its life. Also, all female

mammals have glands that can produce milk. Mammals are among the most intelligent of all living creatures.

Q) If chimps and Humans share 98.8 percent their DNA, why are we so different?
A) Although humans and chimps have many identical genes; they often use them in different ways. The gene's activity, or expression, can be turned up or down as the volume on the radio. So the same gene can be turned up high in humans, but very low in chimps.

Q) What was the worldview of human origins pre-science?
A) Most of the world excepted the biblical account that humans came from Adam and Eve.

Q) What land mass did the biblical writers put the origins of humans?
A) The Garden of Eden located in Asia
Q) What does modern science say about the origins of all human beings on Planet Earth?

A) Modern science says that all humans (no matter where they live today) that walk upright on planet earth came from the sexual relationships between the Black African Woman and Black African Men in Africa.

Q) Do human beings come from monkeys?
A) No there is no credible scientific source, which supports the idea that humans come from monkeys.

Q) Did Charles Darwin teach humans came from monkeys?
A) No Charles Darwin did not teach that we came from Monkeys, he stated that extinct apes where our closes relatives.

Q) Where is the Omo-Valley located?
A) It was located in south-western Ethiopia.

Q) Where were the oldest homo-sapiens sapiens bones found?
A) The oldest homo sapien sapiens bones were found in the Omo Valley.

Q) Who discovered Lucy "Dinkinesh" of Ethiopia?
A) A team of Africans led by Richard Leakey in 1967 and 1974 at the Omo Kibish sites near the Omo River, in Omo National Park in south-western Ethiopia.

Q) How old are these bones in the Omo-valley?
A) The geological layers around the fossils were dated to be about 195 ± 5 ka [thousand years ago]. As of now, this makes Ethiopia the candles of homo sapiens sapiens.

Q) What was the dating process used in dating the bones?
A) The dating process used to date bones is called Argon-argon dating method.

Q) How does Argon-argon dating method work?
A) Argon-argon dating works because potassium-40 decays to argon-40 with a known decay constant.

Bibliography

1. "Online Etymology Dictionary." Online Etymology Dictionary. N.p., n.d. Web. March 30, 2013
2. George G.M. James. Stolen Legacy. New Jersey. Africa World Press, Inc. 1992, p. 9-10
3. Mark, Joshua. "Thales of Miletus." Ancient History Encyclopedia. N.p., n.d. Web. 30 Mar. 2013
4. O'Grady, Patricia. Internet Encyclopedia of Philosophy. N.p., n.d. Web. 30 Mar. 2013.
5. George G.M. James. Stolen Legacy. New Jersey. Africa World Press, Inc. 1992, p.29
6. Dr. Yosef Ben Jochannan. African Origins of the Major Religions, p. XV
7. Milo Rigaud. Secrets of Voodoo. New York. Arco Publishing.1969, p.8
8. Imhotep, Asar. "African Origins of the Word God." Blog post. Http://www.asarimhotep.com/documentdownloads/AfricanOriginsoftheWordGod.pdf. N.p., n.d. Web. 30 Mar. 2013.
9. Strong's Concordance, h8064 (abode of the stars, the sky, the atmosphere)
10. Foley, Jim. "Hominid Species." Hominid Species. N.p., n.d. Web. 05 Dec. 2016.
11. "Understanding Evolution." Understanding Evolution. N.p., n.d. Web. 05 Dec. 2016.
12. Smithsonian's National Museum of Natural History. "Human Fossils | The Smithsonian Institution's Human Origins Program." Human Fossils | The Smithsonian Institution's Human Origins Program. N.p., 01 Mar. 2010. Web. 05 Dec. 2016.
13. CooksonHttps://www.facebook.com/BradshawFoundation Clive. "Discovery of Earliest Homo Sapiens Skulls Backs 'Out of Africa' Theory." Bradshaw Foundation. N.p., n.d. Web. 05 Dec. 2016

Chronology of Human Evolution

Made in the USA
Columbia, SC
01 July 2021